Jorge Tsinine

The Role of the Community in Coastal Erosion Management, Ka-Nyaka Island

Jorge Tsinine

The Role of the Community in Coastal Erosion Management, Ka-Nyaka Island

Coastal Erosion Management, Ka-Nyaka Island

ScienciaScripts

Imprint
Any brand names and product names mentioned in this book are subject to trademark, brand or patent protection and are trademarks or registered trademarks of their respective holders. The use of brand names, product names, common names, trade names, product descriptions etc. even without a particular marking in this work is in no way to be construed to mean that such names may be regarded as unrestricted in respect of trademark and brand protection legislation and could thus be used by anyone.

Cover image: www.ingimage.com

This book is a translation from the original published under ISBN 978-3-639-61092-5.

Publisher:
Sciencia Scripts
is a trademark of
Dodo Books Indian Ocean Ltd. and OmniScriptum S.R.L publishing group

120 High Road, East Finchley, London, N2 9ED, United Kingdom
Str. Armeneasca 28/1, office 1, Chisinau MD-2012, Republic of Moldova, Europe
Managing Directors: Ieva Konstantinova, Victoria Ursu
info@omniscriptum.com

Printed at: see last page
ISBN: 978-620-2-78542-6

Contents

Dedication

To my wife, who supported me whenever I needed her support, so that this training could become a reality, to my children (Osvaldo Jorge, Jorge Junior and Hulda Jorge Tsinine) this victory is for you.

Thank you

I thank GOD for life, and I would also like to thank all the teachers at the Faculty of Earth and Environmental Sciences (FCTA), whom I consider family, and in particular my thanks to all the staff at this faculty.

My many thanks also go to my supervisor, Professor Gustavo Sobrinho Dgedge, from whom I learnt a great deal, his patience in teaching me every step and detail in this world of science and knowledge, are teachings that will be useful to me throughout my life, keep it up Professor and may God give you much health, peace and success in everything your soul desires.

A big hug to Professors Sabil Mandala, José Julião, Suzete Buque, Professor António Soluda, Zacarias Ombe and the others who supported my education during the years I spent with you.

Thank you to the Councillor of the Ka-Nyaka municipal district, Mr Araújo, for your support, the information you provided was very important for this research work.

Thank you to engineer Paulo Manguel from the Maputo Municipal Council, who helped me with the district maps, may you continue like this, you are an exemplary official, good service and kind, may God bless you compatriot

I would also like to thank my colleagues from the 2017-2019 Master's programme, for the time we spent together and for sharing valuable knowledge in academic life. And finally, I would like to thank institutions such as the CMCM, in particular the Directorate of Physical Planning, Urbanisation and Environment, the Directorate of Culture and Tourism of the CMCM, the Biology Station of the Eduardo Mondlane University (UEM), the Pedagogical University of Maputo, in particular the Faculty of Earth and Environmental Sciences.

Summary

Environmental problems began to be discussed with greater interest around the 1970s, and have been gaining ground with the development of technologies and the growing human need to occupy and exploit areas sensitive to erosion, and because of their fragility linked to topography, anthropic activities and natural phenomena. The Republic of Mozambique has been creating instruments and norms that regulate certain activities, especially in areas that are sensitive to environmental problems, in line with various initiatives such as the Intergovernmental Conference held in Tsibilisi (USA) in 1977, which initiated a broad global process of Environmental Education, with a view to creating conditions that reorient the production of knowledge based on interdisciplinary methods and the principles of complexity. The study area is the coastal part of the island of Ka-Nyaka, and as you can imagine, the island is exposed to extreme winds, cyclones, the effects of climate change and their consequences. Because of these phenomena, the research area is struggling with environmental problems linked to erosion along its coastline, so the role of the community in erosion management can have a positive effect on combating soil erosion.

Keywords: The role of the community, management, coastal erosion, Ka-Nyaka Island

Introduction

As a country, Mozambique has been implementing policies and legal instruments to regulate and develop certain activities linked to the preservation of the environment, in addition to Law No. 20/97, of 1 October, the Environment Law and the Regulation on the Environmental Impact Assessment Process, approved by Decree No. 45/2004, of 29 September, in its Chapter 1, Article 5, paragraph b). This section describes the state of the environment in the area of influence (physical, biotic, socio-economic and cultural environments). Topography, geology and soils of the reference situation The following aspects should be described: a) Topographical, geological and geomorphological characteristics; b) Types of soil, their characterisation (biochemical parameters), distribution and aptitudes; possible problems of erosion, flooding, salinisation, acidification, compaction, pollution and others; c) Stability of the substrate in relation to construction; d) Areas with potential instability or that could affect or be affected by the activity: slopes subject to landslides, landslides or avalanches; disaggregated substrates; clarified substrate (in limestone regions); seismic areas; areas subject to flooding; etc.

Mozambique is a country located in the southern hemisphere, in the eastern part of Africa, bathed by the Indian Ocean, and due to its geographical location it is exposed to various natural and man-made phenomena, including extreme winds, cyclones, *tsunamis* and, in general, the effects of climate change, The coastal area covers the entire country from the north to the south of Mozambique, subject to a series of natural risks resulting from various types of hazards, including erosion of coastal soils, landslides in the dune system, rising sea levels, environmental degradation of habitats and species that are significant in terms of ecological quality.

In order to respond to these diversities, Mozambique provides (MICOA 2007), Land use plans understood in their broadest sense as (a) Mapping/zoning the distribution of resources at each territorial level (national, provincial, district, city and town); (b) Delimitation of areas for the different uses (tourism, fishing, agriculture, livestock, forestry, wildlife, mining, energy, housing, industry, etc.) and their coordination and reduction of conflicts of use; (c) Definition of the levels and patterns of exploitation, including technologies (permitted/permitted in each area) and their coordination and reduction of conflicts of use.) and their coordination and reduction of conflicts of use; (c) Definition of the levels and patterns of exploitation, including technologies (allowed/not allowed) in each area insofar as the exploitation and use of certain resources that are vital for the survival of users cannot be prevented.

As a way of tackling various environmental problems, the Republic of Mozambique has created various institutions and policies to respond to the challenges posed by climate change, such as MITADER, ANAC, MOZBIO, BIOFUND, FNDS, CENACARTA, AQUA, which are institutions working to ensure the conservation and preservation of biodiversity and various ecosystems throughout the country.

Soil erosion is one of the biggest global problems related to land degradation, with effects occurring both on-site and off-site (LAL; STEWART, 1990; BELASRI; LAKHOUILI, 2016; GUERRA, 2016). Soil erosion is defined as the process of separation, removal, transport and deposition of soil particles, caused by the influence of the sun, rain, wind and water, and can be accelerated by human activity. Among the various human activities, deforestation, uncontrolled burning, inadequate agricultural practices, use and exploitation of land in areas prone to soil erosion stand out (PIMENTEL, 2006; BERTONI; LOMBARDI NETO, 2012;

PRUSKI, 2013, apud MANDALA 2016).

With these guiding instruments in mind, this dissertation analyses the areas affected by soil degradation due to water erosion in the research area, which is exposed to various extreme winds, and as a result there is a record of eroded areas, mainly in areas that are frequented and exploited by man, and as a result, the eroded areas are visible and tend to gain ground, thus creating greater vulnerability to suffer greater environmental damage than is already recorded at the time of the research.

The PARPA is one of the sectoral plans that defines the following environmental priorities:

- Sanitising the environment;
- Land-use planning;
- Preventing soil degradation;
- Management of natural resources, including fire control;
- Legal and institutional aspects - environmental education, compliance with legislation and institutional training;
- Reduction of air, water and soil pollution e;
- Prevention and reduction of the effects of natural disasters

According to (MICOA 2007), the severity of soil erosion in Mozambique is the result of cultivation in areas with mountainous topography, torrential rains, uncontrolled burning and the occupation of land with a high degree of susceptibility to erosion and a low degree of vegetation cover.

In Mozambique, issues linked to climate change and poverty are a challenge that in some cases exceeds the country's economic capacity to cope with its destructive effects, especially in the coastal zone. Although there are laws and regulations in the country that regulate certain activities with regard to the environment, there are also cases of environmental problems at a global level that are natural phenomena.

1.1 Justification

It is of the utmost importance to know how environmental risks are managed on Ka- Nyaka Island, especially coastal erosion, and the involvement of the local community in the management of coastal erosion and ecosystems along the coastal areas of Ka- Nyaka Island.

Failure to recognise the role of the local community in the management of areas sensitive to environmental problems can lead to other conflicts of interest between the community and other sectors interested in contributing to the resolution of environmental problems, which is why it is important to know how erosion, which is one of the environmental problems, is managed.

1.2 Problematisation

Ka-Nyaka Island is in the far south of the country at the entrance to Maputo Bay and is exposed to all natural phenomena. According to history, Ka-Nyaka Island is the result of soil erosion, which culminated in its removal from Machangulo to, until then, Ka-Nyaka Island, a phenomenon that deserves special attention, considering the level of population growth currently taking place on the island and the pressure exerted on natural resources, including areas sensitive to coastal erosion due to socio-economic activities.

The local community's source of livelihood is fishing and, to a large extent, tourism. The areas most pressurised by these activities are the coastal areas. The pier bridge, which is the place where goods and passengers (tourists) embark and disembark, is one of the areas that suffers the most pressure, due to the excessive use of vehicles along the beach, thus causing soil compaction and facilitating the progression of coastal erosion.

And the same pressure is also being exerted on the mangroves. Some tourist operators, with the involvement of the community, are cutting down and using stakes and other materials from the mangroves to build infrastructures of their interest, thus damaging the ecosystems and areas that are sensitive to coastal erosion, which in a way increases the risk of suffering major environmental damage in the very near future.

The following question therefore arises;

Why not include local communities in coastal erosion management processes as a way of guaranteeing their sustainability?

1.3 General objective

Understanding the erosion management process Ka-Nyaka Island

1.4 Specific objectives

> Characterise the study area;

> Describe the erosion process on Ka-Nyka Island; > Characterise the impacts of erosion on Ka-Nyka Island and;

> Explain the stages of erosion management on Ka-Nyka Island.

1.5 Methodology

For this dissertation the author did:

- Survey of physical geographic data,
- Analysing socio-economic activities,
- Analysing land use and utilisation,

Bibliographical and cartographic research, observation techniques, interviews and the collection of historical data on the island, in order to better understand the phenomena being researched and their evolution over time, as a procedure to harmonise with the objectives set.

Following the theory of FONSECA (2002), methodology is the "study of organisation, of the paths to be taken in order to carry out research or a study, or to do science. Etymologically, it means the study of the paths, the instruments used to carry out scientific research". It is therefore essential to consult the literature and other techniques to investigate processes and institutions in order to better assess the management of erosion risks in areas that are vulnerable and exposed to environmental damage

BOLEA (1984) says that choosing a particular method or technique as the most advantageous for environmental impact assessments is not recommended, as there is no ideal method that applies to all types and all phases of the study. The choice of method will depend on the objectives to be achieved, the availability of data, the characteristics of the project and the specifics of the location, as well as the time and financial and technical resources available.

1.6 Techniques used

1.7 Observation

The author went to the research field, observed the research area and took photographic images of the research area, observed maps of the municipal district of the research area and other documents of great value to this work.

Observation is one of the techniques that contributes to research, since the researcher comes into contact with the research field, which allows the application of knowledge in the collection of material that becomes part of academic discussions. "In this way, research contributes to discussions in academia and to the practical application of its results," (FREITAS; MOSCAROLA, 2002).

In addition to observation, interviews are one of the indispensable techniques for gathering information that is essential for fieldwork. Interviews with some fluent residents on the island provided relevant information for the fieldwork.

1.8 Interview

Nine influential people on the island were interviewed (see appendix I, table 2), highlighting government officials and others in the tourist industry, natives of the island and others who carry out activities of paramount importance to the economy of Ilhéus.

According to GIL (2008), an interview can be defined as a technique in which the researcher appears in front of the interviewee and asks them questions, with the aim of obtaining the data that is of interest to the research. The interview is therefore a form of social interaction. More specifically, it is a form of asymmetrical dialogue in which one party seeks to collect data and the other presents itself as a source of information.

The interview is one of the most widely used data collection techniques in the social sciences. Psychologists, sociologists, educators, social workers and practically all other professionals who deal with human problems use this technique, not only for data collection, but also for diagnostic and counselling purposes. As a data collection technique, the interview is very suitable for obtaining information about what people know, believe, expect, feel or want, intend to do, do or have done, as well as their explanations or reasons for the above (SELLTIZ et al., 1967, p. 273, apud GIL, 2008).

1 Theoretical foundation

According to KOHNKE AND FRANZMEIER & apud MANDALA (2016), a good soil is one that is in balance with the characteristics of its natural environment. Any deviation or imbalance from these natural conditions results in soil degradation. In turn, any form of soil degradation also degrades the environment around it

Erosion is one of the environmental problems that began to be discussed with greater interest around the 1970s. It is an environmental problem that has been gaining ground with the development of technologies and the growing human need to occupy and exploit areas that are sensitive to erosion due to their weaknesses linked to topography and natural phenomena.

In 1971 LEINZ & LEONARDO defined erosion as, "the combined effect of all the processes of land degradation, including weathering, transport, the mechanical and chemical action of running water, wind, ice, etc." Narrow sense, "gradual cutting away of solid rock by the action of rivers, winds, glaciers and the sea".

According to MOREIRA (2005), the coastline of Ka-Nyaka Island is retreating at the latitude of Mount Ka-Nyaka by an estimated 0.11m/year and the high beach would have slimmed by 22m, at an annual erosion rate of 0.81m, between 1973 and 1999. The southern coastal portion, at the latitude of Ponta Torres, consists of interdune plains and beaches, from which dunes rise that locally exceed 40m in altitude, with a cover of forest vegetation.

GUERRA, (1972), defines erosion as "Destruction of the protrusions or indentations of the relief, tending towards levelling, in the case of coastlines, inlets, bays and depressions.

In geomorphology, there has already been a certain reaction against the didactic system of separating erosion and sedimentation, since both are integral elements of the erosion cycle. Over time, there has been an apparent evolution in the study of concepts linked to the nature of erosion or even its classification, with authors and institutions providing various ways of understanding one of these environmental problems.

According to DAEE / IPT, (1989). Erosion can be "normal" or geological, which develops under conditions of equilibrium with the intensity of soil formation "accelerated" or anthropogenic, whose intensity is greater than that of soil formation, not allowing its natural recovery.

According to OLIVEIRA et al (1987), this phenomenon of erosion has been causing a heavy burden on society through the degradation of the soil and, consequently, the water, because in addition to irreversible environmental damage, it also produces economic and social losses, reducing agricultural productivity, causing a reduction in the production of electricity and the volume of water for urban supply due to the silting up of reservoirs, as well as a series of inconveniences for the other productive sectors of the economy.

Erosion refers to the impoverishment of its physical properties, such as porosity, permeability, apparent density, structural stability, decreased gas exchange, decreased nutrient movement, decreased water infiltration rates in soils, increased water erosion, among others. A large part of these problems are generated by soil compaction caused by the use of heavy machinery when preparing the soil (LAL; STEWART, 1990 & apud MANDALA 2016).

Soil erosion, especially in areas where man has intervened, is one of the biggest environmental and global problems (see table 1), which are related to the degradation of ecosystems and their biodiversity, and with effects that are registered in communities and other areas. In fact, in many cases environmental problems are not only manifested in the place where they were caused, but spread globally.

Table 1, types of erosion, form and acting agent

Geological erosion	Occurring on the earth's surface under natural conditions, accelerated erosion is the result of an increase in the rate of erosion over geological or normal erosion, the result of environmental imbalance due to human activities. Gross erosion is the total amount of material detached and removed by the action of erosive agents in a given area in a given time.
Linear erosion	Corresponds to Forms of Erosion Caused by Concentrated Surface Runoff
Water erosion	It develops in four stages: channel formation where there is a concentration of runoff; rapid increase in depth and width where the headwaters move upstream; decline in the increase with the start of natural vegetation growth; and eventual stabilisation with the channel located in an equilibrium profile with stable walls and developed vegetation holding the soil.
Rain erosion	This is the type of erosion caused by the action of rainwater. In general, any soil erosion caused by precipitation can be classified as rain erosion, but in areas where the land is less protected by vegetation and other elements, the effects of water action can be felt more intensely.
Erosion River	This type of erosion is caused by river water transforming its course into deeper valleys than its surroundings. In addition, when there is no vegetation on the banks of watercourses, they are eroded by the force of the water, intensifying silting processes and widening drainage basins.
Marine erosion	Caused by the wearing away of coastal rocks and soils by seawater, contributing to the formation of beaches and coastal landscapes such as cliffs.

Source: author with data from <https://brasilescola.uol.com.br/geografia/tipos-erosao.htm>. Accessed on 15 July 2018& (Magalhães, 2001). VII National Symposium on Erosion Control Goiânia (GO), 03 to 06 May 2001

Erosion is categorised according to whether it is caused by wind, water or glaciers. The ways in which the soil is eroded are superficial and subterranean. Surface erosion can occur in different stages

Erosion is one of the environmental problems that must be tackled by an organisation with the support of local communities, so that they can provide all the support necessary to achieve the objectives set. The community living in the place where any project is to be implemented must feel ownership of the project and there must be a commitment to the cause in order to achieve the common goals.

1.2 Evaluation of the erosion process

i)   It can be assessed by quantifying gully and gully density per unit area ($m2$ or $km2$), area affected by erosion also per unit area ($m2$ or $km2$) and the thickness of eroded soil (%) in relation to the uneroded soil profile. Rates can be expressed in tonnes/m2 or $km2/year$. The rates indicate the soil removed and deposited;

LANTIERI et al & apud MANDALA, (2016) ii) Risk of erosion or erosion susceptibility - this procedure assesses potential or future soil loss, which can also be expressed in tonnes/m2 or $km2/year$ depending on specific environmental conditions, using the following parameters: slope, rainfall erosivity, soil erodibility, use and cover of soils and main soil management practices.

According to MAGALHÃES (2001) erosion can be:

j) **Erosion by** impact is the result of the impact energy of the agent against the soil which, in addition to partially disintegrating the natural aggregates, releases the fine particles, projecting them out of the massif

k) **Gravitational erosion**: this type of erosion usually occurs in very steep areas, such as mountain ranges. It consists of the rupture and transport of sediments caused by the action of gravity, with the gradual deposition of rock particles from the highest locations to the lowest points.

l) **Laminar erosion: this** is characterised by uniform and gentle wearing and dragging over the entire area subject to the agent. Organic matter and clay particles are the first portions of the soil to come off, and these are the richest and most nutrient-rich parts for plants. Although it is difficult to observe, it can be seen by the decrease in crop production, the appearance of roots or even marks on the stems of plants where the soil has been washed away.

m) **Wind erosion:** this is caused by the action of the wind, which causes the weathering of rocks and also transports sediment to areas further away from the erosion points. It is usually a slower process than others involving the action of water.

n) **Glacial erosion:** occurs when soils freeze and consequently move in blocks. It also acts by freezing water, which expands and causes changes in the composition and layout of rocks and soils.

1.3 Coastal erosion factors

According to DIAS (1993), there are many factors that induce coastal erosion. Although some of these factors are (or can be considered) natural, most are the direct or indirect result of anthropogenic activities.

a) Rising sea levels;

b) Decrease in the amount of sediment supplied to the coast;

c) Anthropogenic degradation of natural structures;

d) Heavy coastal engineering works, particularly those deployed to defend the coastline.

MAGALHÃES (2001) states that, "On the surface, erosion depends on the action of precipitation and diffuse surface runoff. Rainfall can evaporate, infiltrate or remain on the soil surface. Runoff is a function of the slope of the land and climatic conditions.

The rise in global mean sea level is related to the earth's natural climatological variability and to disturbances induced by human activities DIAS (1993)

Human activities throughout history have managed to achieve high levels of degradation of natural resources, causing great and irreversible damage to the environment, HERINQUE (2012)

The impact of the water breaks up the soil into finer particles that can be carried by the current. Disintegration and downstream loading occur depending on the intensity of precipitation and the cohesion of the soil. The erosive power of water depends on the density and speed of the runoff, the thickness of the sheet of water, the steepness and length of the slope, and the presence of vegetation

According to DIAS (1993), the consequences for the coast of the gradual rise in relative sea level depend on the typological characteristics of the coastal area in question, namely the existence of well-consolidated rock outcrops, the characteristics of sedimentary accumulations, the existence of cliffs, the average slope of the beach, the presence of dune bodies, the frequency of storms, etc.

Coastal erosion is a global phenomenon that occurs essentially on oceanic shores, but also on lakes, and which has been referred to in technical and scientific literature for several decades CHARLIER AND MEYER, (1998)

"A number of elements contribute to the generation of erosion furrows, among them trails, especially cattle trails, and access roads, the concentration of rainwater, sites subjected to improper agricultural management, with the removal of vegetation cover, etc. The main aggressions caused by man stem from: the removal of vegetation cover, inadequate handling of the soil in agriculture, intensive animal husbandry, the opening of ditches, the opening of roads and allotments".

Water contributes various dynamic effects: the soil through impact, surface disintegration through runoff, subsoil disintegration through runoff, underground due to the upper water table; through the transport of detached or disintegrated soil; through the sliding and falling of sandy massifs; through the share of surplus runoff, both surface and underground MAGALHÃES (2001).

1.4 Stages of the erosion process

ULRICH BECK (2006) identifies the society of erosion with a second modernity or reflexive modernity, which emerges with globalisation, individualisation, the gender revolution, underemployment and the spread of global risks. Today's risks are characterised by their consequences, which are generally highly serious, unknown in the long term and cannot be assessed precisely, as is the case with ecological, chemical, nuclear and genetic risks.

JACOB (1998) says that "the issues that environmentalism raises today are closely associated with the need to establish citizenship for the unequal, the emphasis on social rights, the impact of the degradation of living conditions resulting from socio-environmental degradation, especially in large urban centres, and the need to broaden society's assimilation of the reinforcement of practices centred on sustainability through environmental education".

But in the face of this new reality, there is a need for a global reaction in order to curb the serious environmental problems associated with new development trends. Sustainable development requires each of us to make a rational commitment to the use of available resources.

In this new lifestyle reality, the environmental education component cannot be ruled out

According to REIGOTA (1998), environmental education points to pedagogical proposals centred on raising awareness, changing behaviour, developing skills, the ability to evaluate and the participation of students.

The notion of sustainability therefore implies a necessary interrelationship between social justice, quality of life, environmental balance and a break with the current pattern of development (JACOB, 1998).

Looking at the stages of erosion management, which consist of the process of identifying, evaluating and managing events in the face of critical uncertainties. Uncertainties arise from the inability to accurately determine the possibility of a certain event occurring and its accompanying impacts.

According to PULKOWNIK ET AL. (2005) the steps for assessing environmental risk considering the stressor are similar and can be presented as follows:

Stage 1 - consists of the identification of the risk by the parties involved in the organisation, followed by communication and consultation, involving the interested parties both inside and outside the organisation.

Stage 2 - criteria are defined for analysing and mapping the risks already identified by the organisation and involving environmental, social, cultural and technological issues, etc.

Stage 3 - Evaluation of the risks that must be controlled by the organisation, and this stage involves identifying the causes of the risks and their positive as well as negative consequences.

Stage 4 - Risk treatment and monitoring: In this stage, the priorities for treating the risks raised by the organisation are determined. Generally, a risk analysis matrix is used to standardise and facilitate the process of deciding which risks will be treated and prioritised by the organisation, closing the cycle of risk identification, analysis and treatment, monitoring and critical analysis should be carried out consistently throughout the entire risk management process, with the aim of continuously improving the organisation's management system.

The steps for environmental risk assessment considering the stressor are similar and can be presented as follows, according to PULKOWNIK ET AL. (2005), which is in line with the AQR structure:

1-Identifying the problem (potential dangers);

2-Characterisation of the receiver;

3-Toxicity assessment;

4-Evaluation of the exhibition; 5-

Risk characterisation and; 6-Risk assessment.

Table 2, erosion risks in social areas such as: health, safety, environmental and/or ecological.

Environmental/ecological risks	They focus on impacts on the ecosystem and habitats, and can manifest themselves at great distances from the source. Generally, they are characterised by subtle changes and complex interactions between populations, communities and ecosystems (including food chains); - repeated exposures, the effects of which may not manifest themselves for long periods of time (chronic); - cause-effect relationships with high uncertainty.
Risks of Security/Industrial	They focus on aspects of human safety and material loss, essentially within the workspace. They generally have the following characteristics: - they are accidental in nature - they have a low probability of occurrence and a high consequence; - they have high effects in a short space of time (acute); - they have obvious cause-effect relationships
Risk to Health	They focus on human health, essentially outside the workplace or facility. Generally, they are characterised by: - high probability of occurrence and low consequence; - repeated exposure, the effects of which may not manifest themselves for long periods of time (chronic); - difficult to establish cause-effect relationships.

Source: author/Lupércio França Bessegato PUC Minas Institute of Continuing Education

According to MAGALHÃES (2001), erosion is the result of the impact on the soil's physical properties and has an impact on the environment. Inadequate land use planning can be observed when faced with the growing number of houses built in unsuitable locations and without a proper environmental impact study.

Municipalities should focus on drawing up specific legislation, master plans and land use planning, respecting the peculiarities of the environment. The need for more in-depth knowledge of the laws governing the functioning of the physical environment is fundamental in establishing efficient environmental diagnoses.

Public authorities are a determining factor in environmental law. Specific legislation and the creation of plans to correct the problem are necessary, in an efficient prevention policy, and one of the objectives of this work is environmental education.

1.5 Erosion impact

The issue of climate change is often reduced to quantifying absolute or relative sea level rises. In most cases, the emphasis in analysing evolutionary dynamics is centred on historical variations in the level of ocean waters, as well as future projections that predict more or less accentuated rises, albeit in orders of magnitude of millimetres or centimetres on a time scale of 50 years to a century (RIBEIRO 2010).

The impacts of changes in the flows of ecosystem services on the constituents of well-being are complex and involve mutually reinforcing causal relationships, mainly due to the interdependence of the processes that generate ecosystem services and between the dimensions of well-being themselves. Changes in the provisioning ecosystem services, for example, affect all the constituents of individuals' material well-being. However, the adverse effects of changes in the flow of provisioning services can be lessened by socio-economic circumstances (ANDRADE & ROMEIRO 2009).

Some erosion impacts can delay investment in certain areas, due to the climate of uncertainty that comes with it, jeopardising social infrastructure and other community assets along the coastline

The impacts of the risk of erosion are related to "the growing problems related to the impacts of climate change and the degradation of water resources, which show the interdependencies and affectation of ecosystems and biodiversity on a global scale, rapidly reducing the number of species and genetic varieties. And with the loss of stability in the biosphere, climate stability is lost, as is the production of natural resources and also intangible values (aesthetic, landscape, cultural) which, for example, are transformed into material values by tourism, RIBEIRO (2010) ". Soil erosion generates strong environmental impacts and high economic costs due to its effects on agricultural production, infrastructure and water quality (BREETZKE et al., 2013).

RIBEIRO (2010) adds that "the consequences of climate change will be felt at all levels, including in the fisheries sector, since most fish species are very sensitive to small variations in temperature, salinity and water turbidity, among others".

According to the same author, "The projected results for the year 2100 are worrying for the river mouth areas because, in terms of direct impacts, they imply:

a)   Increased natural coastal erosion;

b)   Increased siltation in estuarine and lagoon areas (with a significant reduction in materials exported to the coast and the entry of greater volumes of sand transported in littoral drift with a possible reduction in coastal sediment transit in some sections);

c)   Accentuation of the loss of salt marsh areas (precisely the areas of land where biological productivity is greatest).

Environmental risk represents the degree of harm or damage caused to people and property by the occurrence of a particular hazard (VARNES 1984).

Human occupation of the coastal zone in areas of greater vulnerability is the main factor responsible for erosive phenomena by altering the dynamics of natural processes. As a response, the need arises to build new heavy coastal defence works that often act in emergency situations to protect people and property. Examples include some low and sandy coastal sectors, but also cliff sections. There are various situations of cliff instability which, due to intense use and construction load, have led to the need for corrective and emergency interventions (RIBEIRO 2010).

By interfering in the natural balance between the soil and the environment (removal of vegetation), man often promotes and accelerates the removal of the surface layer, leaving the subsoil (generally of lesser resistance) subject to intense removal of particles, which culminates in the emergence of erosion hotspots.

The race between the demographic explosion culminates in the deterioration of the land, operating in the opposite direction, but with the effects and consequences of the demographic explosion itself, the population pressure exerted on the areas already occupied by them ends up deteriorating more and more rapidly.

1.6 Erosion mitigation measures

There are ways of mitigating some environmental risks, but it is important to emphasise that there are risks of natural origin and others of anthropic origin, but all can be mitigated depending on the goodwill of man, who until then has the intellectual capacity to intervene.

In order to mitigate the risks of erosion in erosion-sensitive areas, there needs to be collaboration between various institutions interested in environmental issues. RIBEIRO (2010) also advocates a social balance in order to achieve the common objectives, according to him, "ensuring social and territorial balance and a balanced distribution of resources and opportunities among the various social groups, generational classes and territories", can go some way to achieving the recommended objectives.

Given the high level of tourist demand and urbanisation that the area under study is subject to, factors that contribute to increased erosion and the weakening of areas that need special attention, due to their ecological and topographical fragility, with a greater focus on the dunes, the significant retreat that is taking place could in a way increase the loss of sand from the beach, and accelerate the destruction of erosion-sensitive areas such as unconsolidated dunes and vegetation.

Table 3, Mitigation Measures

1	Intervening in risk areas associated with natural and/or man-made phenomena by implementing operational programmes that enable critical situations to be mitigated in the short term, based on the definition of priorities;
2	Safeguarding vulnerable and at-risk areas through the operationalisation of contingency plans and adaptive and forward-looking management based on assessment mechanisms that take into account the dynamics of the coastal zone;
3	Promoting cost-benefit analysis, by making it compulsory for all interventions subject to environmental impact assessment and also in the situations provided for in territorial management instruments;
4	Unified coordination of the specialised emergency intervention bodies, through co-responsibility on the part of the competent bodies, specific ongoing training actions and the adaptation of human and operational resources.

Source: author based on data, (RIBEIRO 2010).

Therefore, "erosion management must be shared by all the organisation's agents as organisational forms. This form of management makes it possible to share responsibility between all members of the organisation.

Urbanisation, the most drastic form of land use, imposes the adoption of poorly permeable structures, resulting in a decrease in infiltration and an increase in the quantity and speed of surface water run-off. Accelerated erosion (anthropogenic action) can be laminar or sheet erosion, when caused by the diffuse flow of rainwater resulting in the progressive removal of the soil's surface horizons; and linear erosion, when caused by the concentration of the flow lines of surface run-off water, resulting in incisions in the surface of the land in the form of furrows, gullies and gullies (OLIVEIRA, 1994).

Monitoring is a way of controlling erosive processes, since maintenance recommendations and measures can guarantee continued functionality in reducing erosive activity. If monitoring is carried out by inspecting erosion sites, with access by land, there are two main terms: stability of erosion sites and recommendations for maintenance work.

Mathematical modelling is widely used to predict erosion, both in conservation planning and in erosion control. The main advantage of applying models lies in the possibility of studying various scenarios at low cost and quickly.

The USLE is an empirical equation used to estimate soil loss by erosion for conservation purposes, evaluated by factors, considering input parameters such as climate, soil, topography and land use and management (PRUSKI, 2013). The universal soil loss equation is expressed by the following formula:

$$A = R * K * L * S * C * P$$

A = corresponds to the soil loss calculated per unit area, (t/ha/year), according to rainfall parameters, management and cultivation plans and erosion control practices; R = is the rainfall factor characterised by the erosion rate caused by rainfall, (MJ/ha.mm/ha);

K = is the soil erodibility factor, i.e. the erosion intensity per unit index of rainfall erosion, for a specific soil that is kept continuously without cover, but undergoing normal cultural operations on a slope of 9% and a ramp length of 25 metres, t/ha (MJ/ha.mm/ha). Refers to the risk that a given soil has of eroding, taking into account the inherent properties of each soil;

L = slope length factor is the ratio of soil losses on any given slope length to the slope length of 25 metres for the same soil and slope grade;

S = slope degree factor: this is the degree of slope in per cent. Relationship between soil losses between any given slope and a slope of 9 per cent for the same soil and slope length; The L and S factors are combined using an equation to form the topographic factor (FT).

C = is the land use and management factor that corresponds to the ratio between soil losses from land cultivated under given conditions and the corresponding losses from land kept continuously bare and cultivated;

P = Corresponds to the conservation practice factor, which is the ratio between soil losses from land cultivated with a given conservation practice and losses when planting in the direction of the slope.

Each factor was introduced into the model to represent critical processes that can affect soil loss on a given slope. The R, K, L and S factors are dependent on natural conditions and the C and P factors are related to the form of occupation and land use, so they are anthropogenic factors (BERTONI; LOMBARDI NETO, 2012; PRUSKI, 2013, apude MANDALA 2016).

To be successful, the application of erosion prediction models in planning or control must take into account important criteria regarding their applicability (erosion on slopes, gullies, etc.), the structure of the model, the data available and the cost of obtaining it, and the prediction of the amount eroded, among other factors For soil conservation to be successful, it is important not to use just one conservation practice, but to combine mechanical, vegetative

and soil practices, as they all complement each other.
Soil conservation must be seen by farmers, extension workers, researchers, authorities and society as part of a greater challenge that we face in this new century: the search for sustainable agriculture, i.e. the production of food in quantity and quality to meet the needs of humanity, without degrading natural resources such as soil, water, forests and fauna.
Www.fazendasmt.com.br

Chapter II

1 Location of the study area

Ka-Nyaka Island is located in the south of Mozambique, specifically in the municipality of Maputo. According to the Directorate General of Geology (DGG, 1997), Ka-Nyaka is an island situated at the entrance to Maputo Bay, in the south of Mozambique, with geographical coordinates of 26°S latitude and 33°E longitude. It covers an area of 42 km^2 and has north-south dimensions of 12.5 km (between Ponta Mazondue to the north and Ponta Torres to the south) and east-west dimensions of 7 km. It is located 32 kilometres east of the city of Maputo, of which it is an administrative part, forming a municipal district, Ka-Nyaka

Map 1, geographical location of the Ka-Nyaka municipal district.

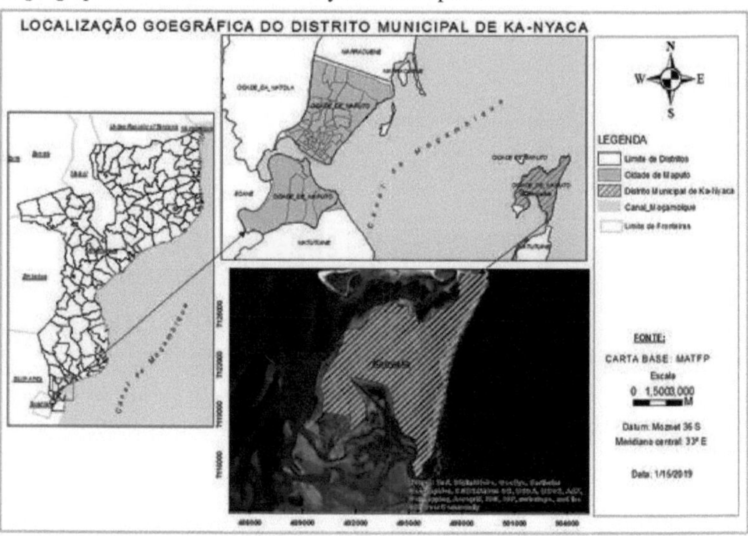

Source: Daniel Sulemane 2019

1.1 Characterisation of the study area

The Municipal District of Ka-Nyaka has three neighbourhoods: Ingwane, Ribjene and Nhaquene. Since its geological origin, Ka-Nyaka Island has been a continuation of the Machangulo peninsula. It is believed that, over time, the land was devastated by erosion, creating the division between the peninsula and the island, now called "Hells Gate" by tourists.

Ka-Nyaka Island has an estimated population of 6,095 inhabitants, according to the latest census, a significant increase compared to the last three censuses. With rapid demographic growth, this prompts us to reflect on the pressure that natural resources are under to meet human needs (Census 2017).

The first inhabitants of Ka-Nyaka were the Tsongas, a Bantu people who populated the coast of Maputo Bay. They are said to have entered via the Machangulo Peninsula to the south. Today, most of the inhabitants descended from the Tsongas are Ronga speakers. However, in 1970, Inyaca Island had less than 1,000 inhabitants, MOREIRA (2005).

In the 1980s, Ka-Nyaka Island already registered a high number of non-native residents, due to the destabilisation war that involved the government and Renamo forces. The same observation is shared by MOREIRA (2005). "In the 1980s, when the island served as a refuge from the civil war, it was home to around 10,000 people".

1.2 Geology

The geology of Ka-Nyaka Island is not dissimilar to the geology of the whole of southern Mozambique, an idea shared in the findings of SENVANO (1997): the geological formations that occur on Ka-Nyaka date from the Middle-Upper Pleistocene to the Recent Pleistocene. They consist of several generations of dunes and coastal marine deposits. These formations have established an undulating relief, with convex forms made up of ancient and recent dunes - most of the ancient dunes are fixed and some of the recent dunes are being remobilised - and concave forms consisting of inter-dune valleys, tidal palaeoplains, interdune plains and sandbars.

The ancient dunes rise on the western side of Ka-Nyaka Island to an altitude of 66 metres at Alto Pocuane and extend predominantly in an SSW-NNE direction, similar to the layout of the dunes on the eastern side, from Ponta Mazonduè to Ponta Torres. The latter, however, are higher, with the dune that makes up Mount Ka-Nyaka standing out at 115 metres high. The central area is an undulating surface, with maximum heights of around 35 metres, and the northern part is made up of tidal and supradunes plains SÉNVANOO (1997).

Ka-Nyaka Island is the result of the erosion of the soils that connected the island with the Machangulo Peninsula. The story goes that the soils were degraded by wind and sea erosion, the soils were transported to other areas and this is how Ka-Nyaka Island came to be, but it sits on the same plate and is considered to be part of the gateway to hell due to its transition from the island to the peninsula.

Figure 1. Dune in the eastern part of the island

Source: author 2016

1.3 Relief

Ka-Nyaka Island has an undulating terrain with consolidated dunes and others in the consolidation phase. Inland, it has plains along its length, and the coastal part has eroded areas. MOREIRA (2005) says that the coastal part of Ka-Nyaka Island is retreating by

0.11m/year and the high beach would have slimmed by 22m, at an annual erosion rate of 0.81m, between 1973 and 1999. The southern coastal portion, at the latitude of Ponta Torres, consists of intertidal plains and beaches, from which dunes rise that locally exceed 40 metres in height, with a cover of forest vegetation.

The floodplain has vegetation with various types of species such as mangroves, small massif trees and typical lowland shrubs, as well as various types of grass that cover the ground along the floodplain, which is completely green in the vicinity of the aerodrome in the southern area of Ka-Nyaka Island.

Figure: 2, completely dry floodplain

Source: author 2016

1.4 Weather

The climate of Ka-Nyaka is moderate humid tropical, with two distinct seasons, a climate characteristic of the southern part of the country, with two seasons predominating, according to SÉNVANO (1997): "the cool, dry season that runs from April to September and the other, hot and humid, from October to March. January and February are the warmest and wettest months, with temperatures of 26.3°C and 26.2°C and rainfall of 135.9 mm and 143.9 mm respectively."

1.5 Hydrology

The lowest monthly rainfall is 23.7 mm and occurs during August, while the lowest average monthly temperature is 19.6°C and occurs in July. The prevailing wind direction is south-west, but from October to January the north-east direction prevails. The highest relative humidity is 82.1% and evaporation, which varies moderately, averages 100mm/month (Mavume, 2000). The warm current of the Mozambique Channel, which bathes the island, and the island's insularity are the main factors responsible for the relative increase in humidity SÉNVANO (1997).

The island does not have any rivers or freshwater streams in the style characteristic of the islands. The story goes that there are "only small lake formations and relatively extensive wetlands (lagoons), the drainage on Ka-Nyaka is only internal, losing moisture through evapotranspiration and gaining it with precipitation. This water circulation is linked to the sea from which the Ka-Nyaka mainland is influenced, and is largely defined by geo-hydrological aspects. The same scenario is also shared by (CNPF, 1990:27).

During the fieldwork it became clear that the communities living on the island consume water from boreholes. All three of the island's neighbourhoods, namely Inguane, Nyaquene and

Ribjene, have boreholes that benefit more than 6,000 inhabitants, while the main town has a small water system that supplies the communities and tourist establishments.

Figure: 3, rainwater mixes with sea water at this low point

Source: author 2019

1.6 Vegetation

The vegetation remains intact, especially in the areas protected by law. There are a variety of species, including chamfer trees, massal trees and many climbing plants, which makes the vegetation areas compact and protected. In addition to the dense vegetation, there are species imported from other parts of the country, particularly coconut palms, acacias, etc.

Vegetation in the prevention of environmental problems involves plants and all their biodiversity, and for its best management a municipal environmental management plan should be in place, which aims to establish guidelines aimed at improving life and its biodiversity in the development of the municipality with the duty to maintain and preserve the environment, above all in the area of its environmental jurisdiction.

In cultivated and inhabited areas, the vegetation is made up of small species, mainly tubers and non-tubers, such as coconut palms, mandioque trees, maize, peanuts, and other derivatives resulting from agricultural activities, mainly for family sustenance, on the shores of the island there are areas that have been uncovered by various factors, mainly the natural factor, through the action of extreme winds and sea waves on high tide days, with the action of the winds removing the soil and transporting it to other areas, thus causing physical soil erosion, and this action is accompanied by marine erosion.

Figure: 4, vegetation mixed with species imported from other areas of Mozambique

Source: author 2016

Figure 5, vegetation cover along the island

Source: author 2016

In addition to being a stable vegetation throughout the year, it creates conditions for housing and breeding birds that nest safely and without human interference throughout the area covered by mangrove species, it is also these areas that have a great potential for corals and reefs, very rich areas of forest and wildlife species, areas protected by law and other regulations in order to protect species and areas sensitive to influences from actions harmful to ecosystems and the variety of their biodiversity.

Figure 6, part of the native vegetation with signs of marine erosion

Source: author 2016

The forested areas along the coastal zone of Ka-Nyaka predominate on the dunes of the east coast and near the Red Barriers in the western part of the island. The same view is shared by QuickBird (2012), the occurrence of coastal forests is distinguished by their deep green colour (see figure 5), without much influence from human activities

Figure 7, natural forest in the island's dunes, with no notable human influence

Covering relatively large areas, but dotted with small patches of pale green. The strip from the southern end of Ponta Torres to the Inhaca Lighthouse, as shared by Kalk (1995) "and observations updated in 2009, forms one of the few portions of the island with a forest ecosystem of low disturbance by human activities. Forests are characterised by having scattered trees separated from each other by extensive areas of grass associated or not with small shrubs (Kalk, ibid). In contrast, the pioneer vegetation forms a carpet of succulent herbaceous species, occurring immediately above the high tide line, on loose sand particles, where the soils are low in organic matter, dry, highly saline and with strong ocean winds".

Map 2 - Cartogram of human interference hotspots

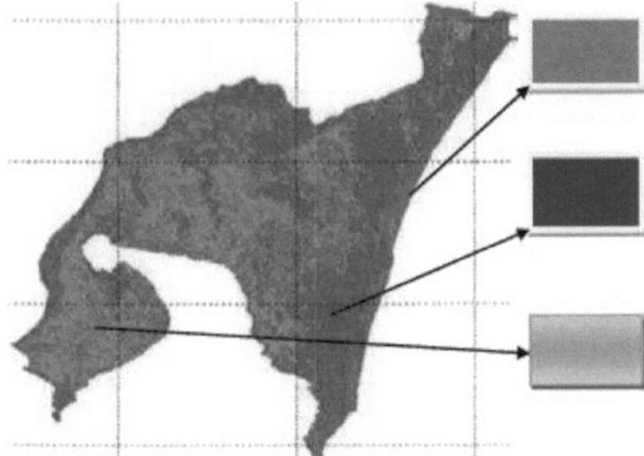

Source: adapted from the 2016 Birth map

Coastal areas vulnerable to erosion Areas with virgin forest Areas with little vegetation

The existing vegetation on the island is natural and much of it is free from human influence due to some awareness-raising work carried out by civil society and other environmentally friendly organisations. These are native forests that contribute in some way to halting the erosion processes around the island.

And the forest plays an important role in the natural maintenance of existing ecosystems, especially in the coastal zone. Thanks to the collaboration of the community, it is still possible to see native species, but community participation is not significant due to a lack of clear guidance. Throughout the island, there is no commercial activity involving woody material, which in a way contributes to the maintenance of the forest on the island; in fact, the community only uses this resource for home use and not for commercial purposes.

Figure 8, partial view of the forest and in good condition

Source: author 2016

In other areas, there are species that have been introduced mainly for tourist exploitation, such as acacias chanfutas, etc. There are areas of mangroves along the coast that have been devastated by residents, some of which have been cut down to decorate tourist infrastructures, while other areas of mangroves have been cut down to build housing for local communities, which are in fact considerable areas.

The pioneer vegetation forms a carpet of succulent herbaceous species, where the soils are low in organic matter, dry, highly saline and have strong ocean winds.

Figure 9, vegetation brought from outside the island

The forest of Ka-Nyaka Island is largely free from human influence, as it is of

difficult to access and its location, but there are also no practices such as deforestation for charcoal production as happens elsewhere in the country, here on the island there is none, nor is there any focus on the sale of firewood, which contributes to the prevention and conservation of the ecosystems that exist there

On the other hand, the vegetation in dune areas is closed and difficult to access. It has a specific and typical composition. The existing vegetation helps to protect areas of the dunes through the vegetation that covers the part of the ground that could be altered by the strong winds that are felt throughout the island and which predominate at the extremities.

Figure 10. Dunes covered in virgin vegetation

Source: author 2016

1.7 Fauna

As far as fauna is concerned, the island has small species, mostly bush pigs, various types of rodents, various species of reptiles and birds, a range of marine species of all kinds such as turtles of various kinds, dolphins, starfish and so on.

Still on the subject of birds, Ka-Nyaka Island has a large number of migratory birds such as the Indian crow, a species of bird that has been causing many environmental problems, due to its behaviour, its somewhat different diet from the type of birds Mozambique has, it has a capacity for production in a short time, even to the point of causing environmental imbalance, it should be noted that this type of Indian crow is a predator of other birds for its diet.

The number of Indian crows on the island of Ka-Nyaka is increasing, and they are currently a major nuisance for locals and tourists. This is a species of crow with special characteristics and unusual habits; according to residents, "it can turn on a tap". As part of the efforts to control the Indian crow, around 6,000 of the animal's eggs have been collected, thanks to support from the Ka-Nyaka Hotel, which has made 30 million meticais available to pay the local population 2,500 meticais for each egg presented and 5,000 for each crow slaughtered. (Notícias, 08/02/01)

Species throughout Mozambican territory vary and depend on multiplication and dissemination factors, including geographical barriers (ocean, lakes, rivers, mountains); climatic factors (temperature, humidity and rainfall); biotic factors (relationship between different living beings); substrate and soil (composition and nature of soils); historical factors (evolution of species) and human activities, which can contribute to their enrichment or impoverishment or total extinction (MUCHANGOS 1999) **1.8 Marine Resources**

According to history, the first scientific exploration of the area by Europeans took place in 1909. Since 1921, lecturers and students from the University of the Witwatersrand have come to the island to study its biodiversity.

Source: author 2016

And it is important to note that the EBM is being managed by the Eduardo Mondlane University for various scientific research activities, as well as work linked to raising awareness among communities in order to preserve some endangered species. They are also working in coordination with various sectors in order to supervise some activities of the local community that fishes along the island so as not to destroy the corals and other areas of conservation and protection of species protected by law in the Republic of Mozambique.

It should be noted that this station is largely dedicated to research work, under the responsibility of the Eduardo Mondlane University. It is clear that in recent times, and with the problems of coastal erosion, more work needs to be done to raise awareness and help change behaviour in order to reduce the erosion problems visible along the island and especially on the side of the quay bridge.

Figure 23, a group of media professionals visiting the Biology Station

Source: author 2016

In this first phase of exploration, it was concluded that the island had a vast number of species that could be studied and so, in 1947, the creation of the Inyaca Maritime Biology Station (EBMI) was proposed. Between 1948 and 1950 the station was built and in 1951 it was inaugurated by Admiral João Moreira Rato, with its management handed over to the Department of the General Government of Mozambique.

According to history, on 20 May 1963, EBMI was handed over to the Scientific Research Institute of Mozambique. In 1965, the Ka-Nyaka Forest and Marine Reserves were created and handed over to the Ka-Nyaka Marine Biology Station for management.

In addition to researchers from Eduardo Mondlane University, there is a record of researchers coming from other countries to carry out research for various purposes. Among the nationalities that are most interested in researching the species on this island, we can highlight Brazilians, Portuguese, South Africans, etc. Ka-Nyaka Island is a world-class heritage site

Figure 11, a), various marine species that serve as samples for visitors to see.

Source: author 2016

In addition to researchers from Eduardo Mondlane University, there are also some native staff from Ka-Nyaka Island, which is good and serves as an incentive for the local community. Some tour guides are young people from the island and spend part of their lives involved in tourism-related activities.

1.9 Soils

According to KALK (1995), on the island of Ka-Nyaka, the soils are sandy in texture, white in colour and poor in fertility, covering a considerable area of hectares. The soils are deep and very permeable. In the lower parts, under obvious hydric influence, are soils whose genesis was determined by the main influence of low topography and poor drainage.

Figure 12, sandy, white-coloured soils along thealha coastline

Source: author 2016

There are low-lying areas on the island which, due to their characteristics, allow communities to develop some agricultural activities, with the production of certain subsistence crops, but much of the island is quite sandy and has little capacity for agricultural activities.

Figure: 13, residential areas with a mix of vegetable plantations

However, other parts of the soil are quite sandy and do not allow for agricultural activities, such as areas that have no vegetation cover, the soil is exposed to extreme winds, soil erosion and other phenomena that do not allow for the practice of agriculture.

There is a small plain, with wetlands, where the water table is not deep, with grey soils, in the vicinity of the lagoon (see figure 6) the hydromorphic soils with fresh water and that in wetlands, the same is shared by MUACANHIA et al (2009), "the possibility of exchange of fresh water with salty sea water is accentuated during high tides, and for this reason the depth of the fresh water table undergoes mobility and variation in salinity".

1.10 Land use and utilisation

The land is used for various purposes, including housing and agriculture. According to the 2005 municipal report, more than 150 hectares are used for agriculture. The use and coverage of the land is the responsibility of the Maputo Municipal Council (CMCM), which is the institution that regulates the use and reuse of the land. There are parcelled areas for housing, public and private developments, wildlife and marine reserves.

Looking at the coastal part of Ka-Nyaka Island, there is a wear and tear of ecosystems due to anthropogenic activities and natural causes, much of Ka-Nyaka Island shows signs of moderate to high risk of degradation

Figure: 14, improvised road in the middle of mangroves

Source: author, 2019

The lack of participatory management in erosion-sensitive areas and the use of vehicles along the coastal areas of Ka-Nyaka Island, is responsible for the consequent damage to the areas close to the jetty bridge and its surroundings, the fencing that had been done decades ago in order to protect infrastructure and ecosystems in the area in question, with the advance of erosion processes the island is at risk of suffering major environmental damage.

In recent decades, the island of Ka-Nyaka has been hit by a series of natural phenomena with significant impacts on the environment, others of human origin due to their behaviour, which in a way ends up penalising local ecosystems.

Some mangrove areas are being devastated by man for various purposes, most notably the construction of some infrastructures using local materials, as mentioned above. Another problem has to do with the growth of some practices that are not recommended for the environment, especially in conservation areas and/or sensitive areas due to their nature, which consist of opening up roads in areas that are sensitive to erosion, such as dunes and areas with little vegetation.

In the last five years, marine and/or coastal erosion has grown considerably, and all because of some activities that are carried out on the beach, and in inappropriate places. Commercial exchanges are made on the beach, areas that are sensitive and vulnerable to erosion due to the large number of people and vehicles that are used there to transport goods and tourists.

The number of vehicles that are used along the beaches to transport goods and tourists has contributed to environmental degradation and, by law, we know that the use of vehicles on beaches is not allowed, as we can see in this study area, since these vehicles are responsible for compacting the coastal soils and consequently create conditions for the advance of sea water that carries away part of the existing sealing.

1.11 Transport

Ka-Nyaka Island has a passenger transport boat named after the island called "Ka-Nyaka". This boat replaced an old one called Nyelete, which was no longer able to guarantee safe journeys due to successive breakdowns, which in some cases ended up in the postponement of journeys. The Ka-Nyaka boat is currently the one that guarantees the transport of people and goods, which was recently inaugurated and handed over by the President of the Republic Filipe Jacinto Nyusi.

Figure: 15 maritime transport link between Maputo and Ka-Nyaka Island.

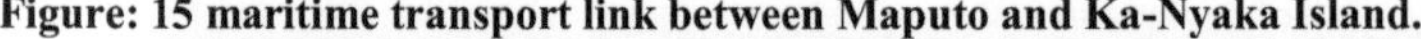

Source: author 2016

In addition to sea transport, the island can be reached by other means such as air transport, but there are also other private boats that operate on this route organised every weekend, and on the other hand there is an ambulance boat at the service of the district hospital, to respond to cases of patients who, due to the seriousness of the situation, need to be transported to other hospitals in Maputo city.

Chapter III

4 Description of the erosion process on Ka-Nyaka Island

The processes of soil destruction through erosion are responsible for the decline in the quality of the environment, and are therefore the consequences of human activities on the natural environment. In general, the processes of soil degradation on Ka-Nyaka Island are classified as physical, chemical and biological degradation.

Soil erosion on Ka-Nyaka Island has to do with the impoverishment of its physical properties, such as porosity, permeability, density, structural stability and the decrease in water infiltration rates in the soil, and consequently opens up space for water erosion

With the need to transport fuel to Ka-Nyaka Island and other products, the chemical degradation of the soil resulting from anthropogenic activities has arisen, which has to do with the socio-economic sector and population growth. Chemical degradation includes acidification, base exchange leaching and salinisation.

The areas suffering from soil degradation due to biological erosion on Ka-Nyaka Island are the result of the loss of some microorganisms that play a very important role in soils, by restructuring or recycling nutrients, which also contribute to the decomposition of plant and animal waste in the environment and thus purifying toxic elements from the environment.

All the factors that reduce or prevent the activities of these important microorganisms in the soil contribute to the biological degradation of the soils in the study area,

Biological degradation is when a certain group of microorganisms existing in the soil is hindered or eliminated, which will often result in changes to biogeochemical processes and the associated media in the environment

Figure: 16, a), b), c) and d), water erosion in sensitive areas, influenced by various activities harmful to the health of existing ecosystems along the beach of KaNyaka Island.

Source: author 2019

4.1 Characterisation of erosion impacts on Ka-Nyka Island;

- Destruction of infrastructure;
- Destruction of ecosystems along the areas affected by erosion;
- Contamination of ecosystems through fuels and their derivatives;
- Disappearance of some species of fauna and flora;□ Threats to biodiversity through the destruction of its habitat.

Many infrastructures are threatened by marine erosion and partly result in the destruction of ecosystems and their biodiversity

Figure: 18, a), b) and c) destruction of seals and water contamination by fuels

Source: author 2016

4.2 The stages of erosion management on Ka-Nyaka Island.

Stage 1 - areas with deficiencies in the management of environmental damage caused by soil erosion were identified, including marine erosion caused by high waves on high tide days, wind erosion and other erosive damage resulting from anthropogenic activities (see figure 18, a), b) and c)) in areas susceptible to soil erosion.

Stage 2 - this phase analyses the seriousness of each situation and the measures to be taken, which may consist of mapping the risks already identified by the organisation and involving environmental, social, cultural, technological, etc. issues, with the aim of prioritising possible emergency situations

Source: author 2019

Stage 3 - Assessment of the risks that should be managed and controlled by the organisation, and this stage involves mapping the causes of the risks and their positive as well as negative consequences and working to mitigate their impacts with preventive and/or corrective measures.

The measures to prevent or correct erosion are divided into 2 groups, to be emphasised in JACOBI's vision (1998), the Treaty on Environmental Education for Sustainable Societies and Global Responsibility sets out principles and an action plan for environmental educators, establishing a relationship between public environmental education policies and sustainability. It emphasises participatory processes in the promotion of the environment, aimed at its recovery, conservation and improvement, as well as improving the quality of life.

a) Preventive measures consist of prior planning for any activity linked to land use, especially urban and rural drainage systems. Problems arise due to the increasing use of land for agricultural purposes. This leads to problems with deforestation, the laying of unlined roads and river drainage systems.

b) Corrective measures include dissipation basins upstream of the start of the gullies and the construction of dams.

Erosion can be contained and controlled by flow, slope level or the nature of the soil or terrain. Controlling the flow is done by diverting the water along predefined paths in relation to the erosion furrow, and controlling the slope is achieved by placing barriers that reduce the speed of runoff, for better control of the land or soil is in the transformation of the surface vegetation cover to make it more resistant to exposure to erosion processes.

Ka-Nyaka Island is not exempt from these damages of varying magnitudes, but its future will depend on the attitudes taken at the present time, whether they are taken by the local or central government, including the local community.

Stage 4 - Treatment and monitoring of environmental problems such as erosion, there is currently no major work on root works to combat erosion problems, although there are promises on the part of the municipal government.

Ka-Nyaka Island must preserve all projects that aim to maintain the sustainability of ecosystems and biodiversity, which currently have areas that are sensitive to erosion due to human activities,

Although there is a disused gate that allowed vehicles to pass through a corridor by the pestana lodge to the ramp that gives access to the beach, the community is currently using a passage that passes through an unconsolidated dune in a disorientated manner, putting the environmental problems that have already reached worrying levels at risk of worsening.

Over the last three years, environmental problems have worsened on the island of ka-Nyaka, and there is no evidence of any action being taken to curb the problems already presented in this research. This reduces the efforts made in the past by the local government and its partners in the private sector to ensure the correct use of spaces that give access to the sea.

Alternative passages that cause greater damage to the ecosystems of the island's coastal zone, goods, heavy loads along the beach consisting of fuel, food and a large number of people along the beach. All erosion problems must be accompanied by environmental education programmes, which are born as an educational process that leads to environmental knowledge embodied in the ethical values and political rules of social coexistence and the market, which implies the distributive question between the benefits and losses of appropriating and using nature. It must therefore be directed towards active citizenship, considering their sense of belonging and co-responsibility, which, through collective and organised action, seeks to understand and overcome the structural and conjunctural causes of environmental problems SORRENTINO (2005).

It is important for the community to be involved in maintaining the resources available for social benefit, and it is always good and important when society feels committed to the public cause

Environmental education is a process of recognising values and clarifying concepts, aiming to develop skills and modify attitudes towards the environment, in order to understand and appreciate the interrelationships between human beings, their cultures and their biophysical environments. Environmental education is also related to the practice of decision-making and ethics that lead to the improvement of the quality of life," (TBILISI, GEORGIA 1977).

Figure 20, a) infrastructure in disrepair due to marine erosion

During the course of the fieldwork, a number of influential people on the island were interviewed, most notably the island's natives, who have a lot of influence within the local communities and beyond, also in the areas of tourist activities.

CP1, a businessman in the tourism industry, on the role of the community in the management of coastal erosion "There is no activity that involves the local community. In the past, we as a community used to use the ramp that runs through the middle of the eyrie to avoid environmental problems such as erosion, But over time the management of that establishment became uncomfortable, perhaps because we were passing close to the swimming pool where it was intended for the clients of the house, and because of this we stopped using the ramp and until now we use an alternative route to do our work."

Figure 21, a), b) and c) one of the ramps next to the swimming pool used to be used by the community to pass through.

CP2 young entrepreneur, "as a community and young people living here on the island, we want to take part in various initiatives to combat erosion here and in other areas, but we don't have the support of the local authorities. In the past, to prevent erosion on the bridge, we used a passageway in the eyrie, but as soon as it's closed there's no way to help prevent erosion."

CP3 Director of Tourism at Maputo City Council, asked about community participation in local activities to contribute to the prevention and conservation of erosion-sensitive areas, "At the moment we don't have any activities underway, but we can say that we are working towards community involvement, we have taken tourists on a cruise to visit the island, but there are other players like the UEM Biology Station, ANAC and other organisations that work mainly in conservation areas."

When asked about the role of the community in welcoming tourists and its role in managing erosion and conserving ecosystems, he said: "It's really in our interest for the community to participate, but perhaps the district councillor could have more input and I in particular don't have much information about the community's role in managing the beach in particular, in order to stem the tide of vehicles along the beach."

CP4 young sculptor, "We are young people who work in the craft area, we manage to sell our products when the cruise arrives with the tourists, apart from that I don't do anything else, it's like that here, nothing is done, everything is at a standstill, sometimes I go to Maputo to visit my family and then I come back, in the past we had a bit of movement with some tourists who came to stay here at the Pestana Loudge, but they have already closed and our business depends a lot on tourists, but we hope that one day things will get better and we will be able to sell our works again."

All the sources that gave their opinion on the current situation in terms of their contribution to erosion management on the island, and the conservation and prevention of biodiversity, were willing to give their input as part of the local community and interested in collaborating with the institutions carrying out ecosystem prevention projects on the island.

Figure 24 Zoning of areas with special attention, the ban on the use of vehicles along the beach

Source: author 2019

These are signs that in the past there has been work aimed at preserving areas that are sensitive to environmental problems, but due to the lack of involvement of local communities, the signs are ignored and accompanied by a lack of monitoring teams along the areas protected by the signs along the areas indicated.

It should be emphasised that, along the beach, there are visible violations of the rules for the prevention of areas exposed to coastal erosion, even with signs prohibiting it. In some cases, these violations are carried out by civil servants working in the municipal district.

Figure 25, a), b) c) and d), areas damaged by marine erosion along the beach and trails giving access to the area restricted to vehicles

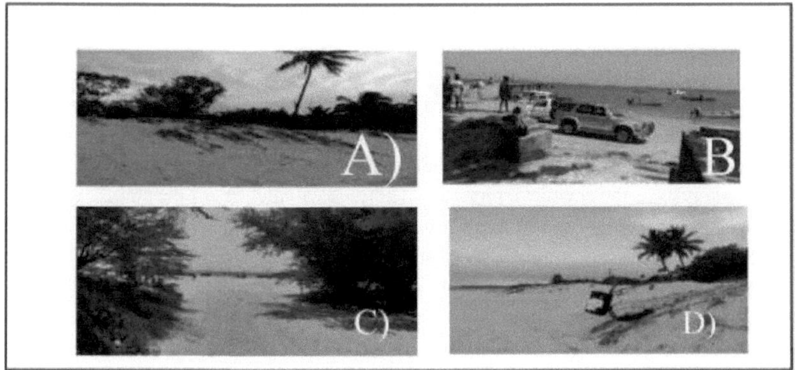

Source: author 2019

In the midst of these vehicles, there are others belonging to the state and the private sector, a reality that until now seems to be normal on this pearl of the Indian Ocean. Next to the quay bridge, there is an office that collects the fees provided for by law for tourists and groups of researchers who come to the island for this purpose.

Erosion control is fundamental to preserving the environment, as the erosion process causes the soil to lose its nutritional properties, making it impossible for vegetation to grow on the affected land and causing serious ecological imbalance and major economic and social losses.

It can be said that of all the natural resources on the planet, soil is one of the most unstable when it is modified, i.e. when its protective layer is removed. Erosion processes occur to a moderate degree in a covered soil, and this erosion is called geological or normal.

According to experts, erosion is a natural and planetary geological phenomenon, without which life would hardly have taken hold on Earth. This phenomenon lowers surfaces, releases elements and makes it possible for organisms to appear.

The problem of erosion leading to desertification becomes serious and worrying when we have accelerated erosion caused by anthropogenic actions, i.e. from outside the environment, those made by man, such as the incorrect use of the soil, without precaution, which results in areas degraded by excessive cultivation, thus going against the recommendations of good agronomic techniques.

Once modified for cultivation or stripped of its original vegetation, erosion begins, capable of removing more material than if the soil were covered.

Soil particles are dragged away by the action of natural factors such as water, wind and waves, which are types of erosion, as well as geological or normal erosion itself, which has the effect of levelling the earth's surface.

It is in the soil that the roots attach themselves so that the plants can grow, at the expense of the water and nutrients available. There is a reciprocal dependence. Without vegetation to protect it, the soil suffers the direct impact of rainfall, whose water runs off the surface causing erosion. They wash away the organic matter that is essential for plant growth and cause furrows that deepen and sometimes end up becoming large gullies.

Soil lost to erosion flows into water sources such as rivers, lakes and estuaries, causing both internal and external effects on agriculture. The internal effects are low soil fertility and the increasing use of correctives. The external damage is the silting up of springs, flooding, difficulties in water treatment, reduced water storage capacity in reservoirs and contamination of rivers, jeopardising fish production.

The direct and indirect consequences of erosion lead to damage that is often irreversible, expressing the loss of soil and water when man began to intensively exploit the soil. For example, when he eliminated the forest, he began to intensively produce commercial crops that offer little protection to the soil, without concern for rational management and additional measures to preserve the chemical, physical and biological integrity of the soil.

The correct use of soils is one of the main items on the syllabus at agricultural schools. When planting crops or forming pastures, it is important that all those who dedicate themselves to using the land for their livelihoods ask for advice from agronomists, who can be hired when production costs allow, or consulted at the Rural Extension Departments run by the public authorities.

Figure: 26 a) and b), gate giving access to the corridor to the beach access ramp, Figure: 27, a) and b), Commercial exchanges on the beach and the empty market.

Source: author 2016

As a strategy for implementing environmental management plans, ARAUJO (2009) suggests the following: (i) mobilising and coordinating partner bodies and institutions, including defining responsibilities and commitments; (ii) establishing a local agenda, with the participation of the different players; (iii) defining the Plan's management system; and (iv) defining instruments for monitoring and updating the Plan, with a view to continuously improving the municipality's environmental aspects.

4.3 Mitigation and Monitoring Stage

S Restore and maintain the ecological integrity of coastal ecosystems;

S Reduce conflicts over the use of natural resources;

S Maintain the health of the environment;

S Facilitate the progress of multi-sectoral development, respecting human values and natural resources, (NRC, 1993; Chua, 1993; Turner & Arger, 1996)

There are signs of human action towards the environment, the planting of coconut palms is aimed at benefiting man but also the environment.

Figure: 28, entrance to the island from the front of the quay bridge, with some infrastructure

Public and a coconut orchard

Source: author 2016

4.4 Legislation applied to areas susceptible to erosion,

Law no. 4/2004 of 17 June defines "the community as a grouping of families and individuals living in a territorial circumstance of locality level or lower, which aims to safeguard common interests through the protection of housing areas, agricultural areas whether cultivated or fallow, forests, cultural protection sites, landscapes, water sources, housing and expansion areas. The community must therefore participate in all projects aimed at developing its areas of jurisdiction."

Involving them in the management of their land or something they own is certainly and always in their interest (see table 4), not that of the visitors, and in this way objectives can be achieved with regard to their involvement in environmental risk management.

Table 4, - Legislative Chronology, Ka-Nyaka Island, According to Law n-10/99 of 7th July

Legal instrument	Effects
Legislative Diploma Number 2375, of 4 May 1963, I Series	Integrates the Ka-Nyaka Marine Biology Station into the Mozambique Scientific Research Institute
Ministerial Order Number 18736, of 5 June 1965, I Series	Determines the delimitation of the lands of the Ka-Nyaka and Portuguese Islands and assigns their management to the Scientific Research Institute of Mozambique
Legislative Diploma Number 2620, of 24 July 1965, I Series	Creates the Ka-Nyaka Island Partial Reserve, among the areas whose management was entrusted to the Scientific Research Institute of Mozambique

Decree Number 30/75, of 23 October, of the Council of Ministers	Integrates the Scientific Research Institute of Mozambique into the administrative structure of the University of Lourenço Marques
Decree Number 12/1995 of 25 April, of the Council of Ministers	Changes the name of the University of Lourenço Marques to Eduardo Mondlane University and Approves the Statutes of Eduardo Mondlane University
XIV Session of the Council of Ministers held on 14 July 2009	Approves the creation of the Ponta do Ouro Partial Marine Reserve, covering the Ka-Nyaka Archipelago

Biodiversity as a variety of living organisms, including genotypes, species and their groupings, terrestrial and aquatic ecosystems and ecological processes existing in a given region.

It is because of the variety of biodiversity that there are protection zones, which are delimited territorial areas, representative of the national natural heritage, set apart for the conservation of biodiversity and fragile ecosystems or animal or plant species.

Also in this Law n-10/99 of 7 July, in the second chapter of article 10, paragraph 5, it states that the management of the protection zones referred to in points a) and b) of n-2 must be carried out in accordance with the management plan drawn up with the participation of local communities and approved by the supervising sector.

Still in the same law, article 13, paragraphs 1 to 4, talks about environmental protection areas, "in order to ensure the protection and preservation of environmental components, as well as the maintenance and improvement of ecosystems of recognised ecological and socio-economic value, the government establishes environmental protection areas that are duly signposted.

Environmental protection areas are subject to classification, conservation and monitoring measures, which must always take into account the need to preserve biodiversity, as well as social, economic, cultural, scientific and landscape values.

The measures referred to in the previous paragraph must include an indication of the activities permitted or prohibited within the protected areas and their surroundings, as well as an indication of the role of local communities in the management of these areas.

Commercial activities are practices whose main objective is to trade, without caring about the environment and even less about the problems that could affect erosion-sensitive areas in the future.

The community of Ka-Nyaka can assess environmental impacts, BOLEA (1984) "studies carried out to identify, predict and interpret, as well as prevent, the environmental consequences or effects that certain actions, plans, programmes or projects may cause problems for health, human well-being and the environment". These studies include alternatives to the action or project and presuppose public participation, representing not a decision-making instrument in itself, but an instrument of knowledge at the service of the decision".

The increasing expansion of activities along coastal areas without taking into account the potential and limitations of these areas is a means of environmental degradation.

The aim of this work is to assess the management of the physical environment of the coastal areas of Ka-Nyaka Island based on the classification of the use capacity of erosion-sensitive areas, mainly in areas with a population cluster, due to the constant need for trade and other activities related to fishing, tourism, transport, etc.

From 2014 to 2018 the degradation of some ecosystems and infrastructures of community interest worsened in terms of environmental damage

The large hotel infrastructure in front of the pier bridge is at serious risk of disappearing if urgent action is not taken. From 2014 to 2017 and in the first half of 2019, the island suffered environmental damage, with a greater focus on coastal erosion, and the inappropriate use of sensitive areas accelerated the destruction of the protective structure of the infrastructure areas (see images in figures 17 and 18).

The damage is enormous; in fact, these problems are the result of misuse of areas that are sensitive to coastal erosion or damage to ecosystems in the vicinity of where commercial activities are carried out, or other behaviours that threaten the biodiversity that exists in the area.

As can be seen from the whole process of environmental degradation, man appears to be the main cause due to his environmentally unfriendly actions (see image 22). There was protection as a measure to mitigate the damage that in a way already constituted a danger in this area, but as there have been no changes in the behaviour of the users of these areas, the scenario is regrettable.

Figure: 30 e a), The coastal protection wall before it collapsed and a road leading down to the beach.

Source: author, 2019

In order to respond to some of the concerns that afflict the communities living on Ka-Nyaka Island and in areas of environmental risk by involving the community in the fight against environmental problems, there is an urgent need for government intervention to minimise the environmental damage visible in the research area.

Managing environmental risks is an extremely important task for our country, given its geographical location. It is impossible to stop environmental damage, especially along our coast, but each situation can be assessed and managed in the best way to reduce its environmental impact locally and for other areas that may be affected.

With industrial development, environmental problems tend to grow, a view shared by FRANCISCO TAUACALE (2010). With the accelerated development of industrialised countries, especially in the post-war period, serious problems of environmental and/or social degradation became evident. The increased awareness of the populations involved led to a growth in demands for better standards of environmental quality, in which these concerns were incorporated by the government either through preventive or corrective actions and/or alternatives to the development model adopted.

The Ka-Nyaka community must actively contribute to environmental risk management, but there are a number of activities that deserve their attention.

RENN (1992), "Environmental Sociology", the main currents in the sociology of risk have followed three separate but complementary directions that are brought together through an underlying focus on the social context in which individual and institutional decisions about risks are made.

So there are a series of decisions that need to be taken by government institutions with a view to involving the communities that will give continuity and length to the management of the common heritage, the community can participate in behavioural and social monitoring, in spreading good practice

The Stockholm declaration recommended that governments take action to control the sources of pollution, and the 1970s saw the flourishing of pollution control laws and the emergence of government bodies in charge of environmental monitoring and the inspection of polluting activities.

As part of its environmental responsibilities, MITADER has been carrying out a number of activities aimed at protecting the environment through laws that penalise offenders with fines for those who do not comply, and some activities require an environmental licence before they can be implemented as a way of regulating and reducing environmental damage caused by the careless actions of human beings.

Some projects that deserve attention in this research are those aimed at correcting certain environmental problems, but led by MITADER in terms of funding and their realisation.

4.5 Community adaptation action plans

PACA aims to help increase Mozambique's resilience to the impacts of climate change by implementing concrete adaptation measures, identified through community consultation and participatory budgeting processes. This programme operates in the geographical areas most vulnerable to climate change, in accordance with the Environmental Education, Communication and Dissemination Programme (2009-2025).

It also contributes to the National Strategy for Adaptation and Mitigation of Climate Change, which establishes guidelines for action to create resilience, including the reduction of climate risks, in communities and the national economy and aims to promote low-carbon development and the green economy by integrating them into the sectoral and local planning process.

4.6 Standing forest project

Forests play an important ecological role in conserving water, soil and biodiversity, and are also an important source of raw materials and food products in rural areas. Forests provide a variety of climatic, environmental, economic, social and cultural services. The Standing Forest Project's objectives are as follows:

a) Promote rural development based on the protection, conservation, enhancement, creation and sustainable use of the forest; Support the private sector by developing the national timber industry, diversifying and maximising the forestry sector's value chain;

b) Stimulate job creation in the forestry sector through the diversification of goods and products based on the conservation paradigm;

c) Capitalise on the search for and application of international funds and national revenues in the protection, conservation, enhancement, creation and sustainable use of the forest;

d) Mitigate the impact of restructuring and reform measures in the forestry sector;

e) Promoting the development of local communities through support for community management and the valorisation of forest resources, particularly non-timber products;

f) Creating alternatives to the unbridled exploitation of the forest, through preservation activities and increasing the forest stock.

The Standing Forest Project includes the certification of 50 forest concessions, the training of 800 protection inspectors, the construction of 5 inspection camps, the distribution of 12,000 beehives to 800 beekeepers' associations, the installation of 20 honey processing units, the creation of 20 district carpentry shops and the opening of 20 forestry and afforestation nurseries.

4.7 Safe land project

Land is the most important natural resource for socio-economic development. It is where we live, produce and extract the food that sustains life. Its management, administration and awareness provide greater territorial control, productivity and harmony in the coexistence of communities.

The aim of the Terra Segura project is to strengthen the land administration and management system, with a focus on access, registration and information management.

The demographic explosion and the growing demand for natural resources in which we live are raising new challenges for the state's ability to manage the land. These challenges include:

(i)   The concept of land,

(ii) The idleness of the land,

(iii) Spatial planning e,

(iv) The sustainable exploitation of   land.

4.7 The safe land project will implement:

(i)   The registration and titling of the Right to Use and Enjoy the Land (DUAT),

(ii) The mapping of the territory on a scale of 1:50,000 and 1:25,000 of the national territory,

(iii) The construction of an up-to-date, levelled and transparent national cadastre,

(iv) Mechanisms and clients for issuing DUATs, (v) decentralisation of the technical capacity for land management and administration to the districts,

(v) Publicising the rights and obligations of use and exploitation in the communities (Land Law),

(vi) Optimisation of land use through the transfer of cultivation techniques to increase productivity levels,

(vii) Democratisation of access to land while respecting gender equality and, (viii) Security of tenure.

The Terra Segura Project includes the registration and regulation of the occupation of 5 million plots (DUATs), the conclusion of the process of installing the Integrated Land Management System, the organisation of 4 campaigns to publicise land use rights and obligations, the training of 800 agents to supervise land use regulations, as well as 4 trainings for district administrators on land management.

4.8 Mozbio project and its project description

Conservation areas are the habitat for countless species of fauna and flora with ecological and scientific value, and represent an important tourist economic value for the development of rural communities.

Mozbio, the Biodiversity Conservation Project, has the following objectives:

(i)   Training in the supervision of conservation areas,

(ii) Infrastructure in conservation areas

(iii) Promoting tourism

(iv) Map conservation areas and;

(v) Developing communities close to conservation areas.

In this context, the MozBio project includes the construction of 3 aerodromes, the construction of 5 inspection camps and the training of 800 inspectors

According to MITADER, In Mozambique, rural areas represent 90 per cent of the national territory and are home to around 68 per cent of the country's total population. Created with the mandate to ensure and promote sustainable and fair development, the mission of the Ministry of Land, Environment and Rural Development (MITADER) is centred on reducing socio-economic inequalities, with an emphasis on rural areas, by promoting a diversified and inclusive economy.

The creation of MITADER covers, as its name suggests, a multiple and cross-cutting area of intervention (Land, Environment and Rural Development), creating a unique opportunity to focus the government's priority policies and ensure the implementation of a National Rural Development Programme in an integrated and comprehensive manner.

In this context, MITADER has defined a National Sustainable Development Programme, the PNDS, made up of projects and objectives with concrete and decisive targets to ensure the economic, social, environmental, structural and institutional transformations aspired to in the Government's Five-Year Programme (PQG 2015 - 2019), in the Social Economic Plan (PES). The PNDS also responds to the targets set in the Sustainable Development Goals (2015-2030).

The PNDS was conceived on the principle of integrated development in rural areas, through the sustainable use of natural resources, land organisation and environmental management. The programme aims to foster a basic local economy by complementing the supply of basic services, training and attracting investments that are important for development, while exploiting local capacities and innovating local knowledge.

4.9 Environment in motion project

The conservation of soil, flora, fauna, water and air is vital for the social development of existing rural communities and future generations. In this context, the preservation of the environment requires the sustainable management of resources. The "environment in motion" project prioritises action:

(i)   Environmental education,

(ii) Effective monitoring,

(iii) Construction of solid waste management infrastructures and (iv) construction of other infrastructures.

The Environment in Motion project includes the construction of 10 sanitary landfills, 15 environmental education campaigns (including the importance of environmental sanitation and biodiversity).

4.10 Co-operation strategy

In order to better implement activities aimed at development and cooperation at local community level, strategies should be adopted that see community cooperation as an effective means of achieving the objectives of the community and other organisations with regard to the sustainability of projects to be developed at local community level and in areas of their implementation.

Ka-Nyaka Island, due to its nature and geographical location, is exposed to various natural phenomena, some of which are destructive in nature, such as strong winds, cyclones, tidal waves, etc. It needs effective and comprehensive co-operation strategies for multi-sector coordination to achieve the desired objectives.

In this case, it is the co-operation with the local community, especially in the management of problems related to environmental sustainability, the community of Ka-Nyaka, which can play a very important role in the management of areas vulnerable to coastal erosion, mostly due to the lack of community involvement in the management of areas susceptible to erosion, as there is no activity to protect areas vulnerable to erosion and other environmental damage.

4.11 Rural communication

The programme should be accompanied by a rural communication policy that broadcasts constructive and developmental content through the media, with an emphasis on community radio and television, community leaders and local administrative authorities.

The island has a number of outbreaks of various types of erosion caused by human activity, through mangrove cutting for the construction of infrastructures and other properties of community interest, and for other reasons linked to the opening of access roads to various places of interest without observing environmental issues.

For these and other reasons, the island needs environmentally sound management and the involvement of local communities in order to increase their awareness of environmental issues, which require urgent intervention in order to reduce their exposure. Erosion and high waves have knocked down the protective wall of the Loudge crag to such an extent that if it isn't intervened on urgently it could increase the levels of destruction and put several nearby properties at risk. In this order of destruction, we are facing a very worrying environmental risk

As you can see, the community of Ka-Nyaka Island must be involved in the management of environmental risks within its geographical space, the community of Ka-Nyaka Island is part of society, it is important that they are the first to be concerned about the common good, the sociologist JOHN A. HANNIGAN in his work "Environmental Sociology", says that "risk bearers are victims who bear the direct costs of working and living in dangerous environments. HANNIGAN in his work "Environmental Sociology", says that "the bearers of risk are victims who bear the direct costs of working and living in dangerous environments.

The community living on the island of Ka-Nyaka must take a leading role in the defence of its geographical space and monitor actions that could pose an environmental risk.

At the arrival point of some boats from Maputo and other nearby areas (see figure 20), visitors disembark as well as some local residents, and they bring with them large quantities of goods for the purpose of trading and other construction materials, some behaviour that is not appropriate for this area is visible.

Figure: 31, unloading of various products on the beach of Ka-Nyaka Island

Source: author 2016

We all know that in any area where vehicles circulate, the soil becomes compacted and loses its natural state, making it vulnerable to all kinds of changes resulting from the pressure exerted by vehicles and thus exposed to all kinds of environmental risks.

Figure 32, vehicles and loads along the beach, which has already been severely eroded and compacted.

Figure: 33, various food products in a sea transport, including fuel transported in various sized packages

The causes of coastal erosion on ka-Nyaka Island are attributed to a range of natural factors and various anthropogenic interventions in the coastal zone.

Throughout the fieldwork, it was possible to spot a considerable number of vehicles parked on the beach, a scenario that happens on a continuous basis without anyone connected to the local government or community organisations being able to repudiate it and promote actions aimed at correcting the current local behaviour

According to KLEIN (1998), in order to promote more effective management of impacts in coastal zones, it is necessary to understand how the coastal zone reacts when subjected to a wide range of socio-economic and environmental pressures, which interact on short and long time scales.

4.12 Management involving the local community

According to Law n-19/97 of 1 October, "nature protection zone: a public domain asset, intended for the conservation or preservation of certain animal or plant species, biodiversity, historical, landscape and natural monuments, under a management regime preferably with the participation of local communities, determined by specific legislation.

An environmental management plan involving the community has the advantage of generating products that benefit the community itself. Perhaps the concept of natural resource can shed light on the benefits that the conservation of ecosystems (areas susceptible to marine erosion) brings to man in the Management of natural resources.

It is the set of actions designed to regulate the use, control and protection of natural resources. Its need emerged in scientific and political debates, mainly in the 1960s and 1970s, around the interest and concern of environmental movements, regulations, non-governmental organisations, international organisations, among others, regarding environmental issues and the disorderly and devastating use of natural resources, (CARVALHO; CURI;LIRA, 2013,P 31;LACERDA; CÂNDIDO, 2013, P.13).

4.13 Services Systemic echoes

Ecosystem services are environmental goods and services that people obtain from natural ecosystems. Ecosystem services consist of the processes through which natural ecosystems sustain and satisfy the human population, maintaining biodiversity and producing goods such as pharmaceuticals (DAILY, 1997).

Therefore, environmental conservation and protection is the guarantor of humanity's well-being, and for this and other reasons we must conserve and protect fragile ecosystems (coastal dunes, river basins, mountainous areas, mangroves) and rehabilitate degraded forests and the preservation of the environment is the same as guaranteeing the well-being of human beings and all living beings that depend to some extent on nature.

Community planting also includes the planting of shade, fruit and ornamental trees in public spaces (schools, hospitals, gardens, etc.), backyards, villages, town and city streets for the purpose of beautification, protection from wind and sun and generally to improve the quality of the environment (DNTF 2009).

4.14 Main activities within communities

According to (DNTF 2009). The actions to be carried out to develop community plantations are as follows:

a)   Developing education and awareness campaigns for the culture of planting and caring for trees;

b)   Supporting networks of tree planting organisations at district level;

c)   Integrate content on tree planting, treatment, management and protection into school curricula at all levels;

d)   Developing the culture of planting trees by establishing commemorative dates for trees, forests and nature;

e)   Create a network of district and community forest nurseries to produce and distribute plants at favourable prices and conditions;

f)   Prepare and disseminate manuals on planting, care and protection of trees for shade, wind protection, ornamentation and, in general, trees of multiple-use species;

g)   Through the agricultural extension network, promote agro-forestry practices and technologies to improve fertility, protect soils and recover degraded areas.

4.15 Conservation and environmental protection plantations

According to the DNTF 2009, there are several areas of marked environmental degradation in the country, due to the increase and displacement of coastal dunes, deforestation, forest and mangrove exploitation, mining, shifting agriculture and the operations of large agricultural projects that require rehabilitation.

In addition to supplying forest products, both timber and non-timber, forests also fulfil important environmental functions for the country's sustainable development, particularly protection against natural disasters, in the water and nutrient cycle, water catchment and regulation in river basins and their protection.

Erosion control and soil conservation, maintaining agricultural productivity and, in general, forests, directly or indirectly, contribute to the improvement and well-being of populations, both in the countryside and in cities. Currently, the importance of forests for the well-being of humanity is being emphasised at a global level for their role in sequestering carbon and mitigating the effects of climate change.

Conservation and protection reforestation is the establishment of forest plantations with the main aim of rehabilitating and conserving the environment.

The forestry sector in Mozambique has successfully rehabilitated vast degraded areas by establishing forest plantations such as casuarina plantations in coastal areas. However, many of these plantations are over 80 years old and many trees are dead or dying.

The younger plantations have not been properly maintained and are suffering from successive burn-offs and destruction due to insufficient resources

According to SOUZA, (2009), "Beaches are public goods for the common use of the people, and free and unrestricted access to them and to the sea, in any direction and direction, is always guaranteed, with the exception of stretches considered to be in the interest of national security or included in areas protected by specific legislation.

The Municipal Government, in conjunction with the environmental agency, will ensure access to beaches and the sea within the framework of urban planning."

At the level of our institutions, such as the National Directorate of Land and Forests (DNTF, 2009, pg: 8), there are clear guidelines on how to use sites for the conservation and preservation of sensitive ecosystems in environments that could disturb or jeopardise their integrity.

"In addition to the degraded areas resulting from human action, the country has extensive areas of fragile ecosystems and river basins that deserve special treatment in terms of protection and conservation, such as the dunes along the coastline and other areas subject to erosion. Forest plantations can therefore help to minimise the impacts of degradation in the coastal zone by fixing dunes and stabilising watersheds."

4.16 Scope of intervention in areas sensitive to environmental problems

Conservation plantations will be established along coastal dunes, river basins, mountainous areas, in mangrove swamps and in strategic locations for the protection and conservation of ecosystems and biodiversity. DNTF, (2009, pg 27).

These plantations also include the planting of trees to protect springs and water lines, riverbanks, mountain slopes and other areas with steep slopes, using native and exotic species appropriate to each site.

The role of the community is important in the conservation and preservation of areas susceptible to coastal erosion. The environment needs to be in good shape if it is to continue to provide man with the basic conditions for his survival.

Conclusion

Ka-Nyaka Island is a heritage site with resources of high environmental value, it has marine reserves where various activities are underway, including scientific research, tourist activities of various levels, given its ecological and environmental value, we are all called upon to preserve it.

It is important that the process of planning land use and utilisation is shared at a local level, involving the community in combating all the practices that lead to environmental problems, Ka-Nyaka Island shows traces that in the past work was done to preserve areas sensitive to erosion due to human activities.

Although it is no longer in use, there is a gate that used to allow vehicles to pass through a corridor that leads from the pestana lodge to the ramp that gives access to the sea. Currently, the passage through the corridor is fenced off and in disuse, and the community uses an alternative passage through an unconsolidated dune in a disorientated manner, putting the environmental problems that are already reaching worrying levels at risk of worsening.

In order to combat marine and/or coastal erosion, it is essential to involve the local community in environmental risk management and make them feel responsible for maintaining all the areas under their jurisdiction. Throughout the fieldwork, the existence of degraded areas with the involvement of the locals was noticeable, and in the past they had already implemented some projects aimed at combating environmental problems with a focus on erosion along the coast.

In recent times, these projects have been abandoned and the local community and visitors are misusing sensitive areas that are vulnerable to various phenomena of a natural or man-made nature that can cause undesirable environmental damage, such as: unconsolidated dunes, the opening of paths in unsuitable places, the abusive and unregulated use of vehicles on beaches

Recommendations

Land-use plans need to be drawn up that take environmental issues into account and aim to maintain and sustain them over time, taking into account the natural physical phenomena of high value in ecosystems and their biodiversity, and enabling community integration in their management and occupation.

Strictly prohibit the circulation of vehicles along the beach, to avoid soil compaction caused by the weight of vehicles travelling through these areas, soil compaction causes incalculable damage.

Compaction can affect soil strength in several ways: in pre-compaction, where there will be an increase in soil density, in particle distribution where it will be more uniform, in the volume of water that will increase with the reduction in soil volume, the water content in the soil will also be altered due to pressure and mobility.

This team will be responsible for monitoring all activities along the coast and in sensitive areas exposed to various erosion risks.

We know that unconsolidated dune areas are susceptible to erosion, and depending on how they evolve, this can cause significant damage to the nearby ecosystem.

Bibliography

ARAUJO, E. Construindo o Plano Municipal do Meio Ambiente: Texto de apoio pedagógico para as oficinas do Programa Nacional de Capacitação de Gestores Ambientais - PNC/PR - (Building the Municipal Environmental Plan: Pedagogical support text for the workshops of the National Training Programme for Environmental Managers - PNC/PR).
Government of Paraná,   2009. Available at: <http://www.meioambiente.pr.gov.br/arquivos/File/coea/pncpr/Plano_Municipal_Meio_AMb ien te_EliasAraujo.pdf>. Accessed on: 14 January 2014.

ARAÚJO, M. G. M. De. Maputo City. Contrasting Spaces: from urban to rural. In: 10., 2005, São Paulo. Proceedings... São Paulo: University of São Paulo, 2005.

BOLEA, T. Evaluación dei impacto ambiental. Madrid, Mapfre, 1984.

BRAZIL. Constitution of the Federative Republic of Brazil, of 5th October 1988. Available at: <http://www.planalto.gov.br>. Accessed on: 15 December 2011. Stockholm Declaration on the Human Environment, June 1972. Ministry of the Environment (MMA). Available at: <http://www.mma.gov.br>. Accessed on: 28 June 2012.

BREETZKE, G. D.; KOOMEN, E.; CRITCHLEY, W. R. S. GIS-Assisted Modelling of Soil Erosion in a South African Catchment: Evaluating the USLE and SLEMSA Approach.InTech. 2013. Available at:). Accessed on: 24 August 2016.

Tbilisi intergovernmental conference and environmental education 1977

CHARLIER, R. H., MAYER, C. P. D., 1998. Coastel erosion, response and management. Sringer, Berlin, 343 pp.

CNPF - National Planning Commission. Inhaca Island Integrated Development Plan. National Institute for Physical Planning. Maputo, Mozambique, 1990. 156p.

- http://www.mma.gov.br/educacao-ambiental/politica-de-educacao-ambiental
- http://portal.mec.gov.br/secad/arquivos/pdf/educacaoambiental/tratado.pdf

DANIEL CAIXETA ANDRADE & ADEMAR RIBEIRO ROMEIRO n. 159, May 2009.Natural capital, ecosystem services and the economic system: towards a more sustainable

"Ecosystem Economics

DA-SILVA, P.C; CRIPPS J.C. 2005. Mapping areas at risk of landslides and flooding in residential areas of Diadema (SP). In: Brazilian Congress of Engineering and Environmental Geology, 11, Florianópolis. Proceedings. ABGE, São Paulo, 2005. Proceedings (ISBN 857270- 017-X)... ABGE, CD-ROM: p. 892-907.

DE ENCOSTAS, 4, Salvador (BA), 2005. Proceedings... ABMS, CD-ROM: p. 3-15.

DAEE/IPT, (1989). Department of Water and Electricity. Technological Research Institute of the State of São Paulo. Erosion control. Conceptual and technical bases. Guidelines for urban and regional planning. Guidelines for the control of urban gullies. IPT, São Paulo, SP. 92p.

DIAS J. M. Alverinho (1993), Estudo de Avaliacao da situacao Ambiental e proposta de Medidas de Salvaguarda para a faixa costeira portuguesa (Coastal geology)

OLIVEIRA, A. M. S.. Methodological issues in regional erosion diagnoses: The pioneering experience of the Peixe Paranapanema-SP basin. In: National Symposium on Erosion Control, 1997. Marília. Proceedings... ABGE/DAEE, p. 51 - 71, 1987.

OLIVEIRA. M. A. T, COELHO, NETTO, A. L, AVELAR, A. S. (1994)- morphometry of slopes and development of bacorocas in the middle valley of the paraiba do sul river revista geociências, são Paulo, 13 (1): 9-23p.

ENGELEN, J.V.KAUFFMAN, S. Recognising the soils of Inhaca Island using a physiographic map. Pedology and drainage. Maputo: Faculty of Agronomy - UEM & Institute de Investigação Agronómica de Moçambique - FAO, 1977. 16p

HENRIQUE, Felipe Mendes. Morphopedological analysis applied to understanding water erosion processes on slopes in the municipality of pilões PB. Dissertation (master's degree in geography) - federal university of rio grande do norte, Natal/RN. 2012, 133p.

FONSECA, J. J. S. Metodologia da pesquisa científica. Fortaleza: UEC, 2002. Workbook

FRANCISCO TAUACALE. 2010. Environmental Impact Assessment Methods

FREITAS, Henrique; MOSCAROLA, Jean. From observation to decision: research methods and quantitative and qualitative data analysis. RAE-electrónica, v. 1, n. 1, p. 2-30, jan/jun. 2002.

FLÁVIO Rodrigues do Nascimento evaluation of natural resources on inhaca island (indian ocean, Mozambique): first approach, 2016.

HANNIGAN, John A. Environmental Sociology - the formation of a social perspective. Translator Clara Fonseca. Lisbon: Instituto Piaget, 1995Available at: <http://www.diritto.it/docs/24015o- papel-dos-municios-na-competência- Ambiental>. Accessed on: 30 January 2015.

JACOBI, P. et al. (eds.). Education, environment and citizenship: reflections and experiences. São Paulo
Paulo: SMA, 1998.

JOHN A. HANNIGAN. 1995. Environmental sociology, the formation of a social perspective

KALK, M. A natural history of Inhaca Island, Moçambique: Witwatersrand University Press, 3rd Edition, 1995. 365 p

LAL, R.; STEWART, B. A. Soil Degradation. Advances in Soil Science, Volume 11. New York: Springer - Verlag, 1990. 351 p

LANTIERI, D. Erosion mapping using high - resolution satellite data and Geographic Information System: Pilot Study in the State of Parana, Brazil. Rome: FAO, 1990. 173.

LIMA, J. M. Relationship between erosion, iron content, physical and mineralogical parameters of soils in the Lavras region (MG). 1987. 86 p. Dissertation (Master's degree in soils and plant nutrition) - Lavras Federal University.

MACNAE, W.; Kalk, Margaret The Fauna and Flora of Sand Flats at Inhaca Island, Mozambique. Journal of Animal Ecology, vol. 31, n.1 (February, 1962a), p. 93-128

MACNAE,W.; Kalk, M. The ecology of the mangrove swamps at Inhaca island, Moçambique. Journal of Ecology, vol. 50, no. 1 (February, 1962b), p. 19-34

MAGALHÃES, Ricardo Aguiar. Erosion: Definitions, Types and Forms of Control. VII National Symposium on Erosion Control. Goiânia- GO, 03 to 06 May 2001.

MARCHIORI-FARIA, D.G; FERREIRA, C.J.; ROSSINI-PENTEADO, D.; FERNANDES

MARINA DE ANDRADE MARCONI & EVA MARIA LAKATOS 5ª Edition 2009 scientific methodology, science and scientific knowledge, scientific methods, theory, hypotheses and variables, legal methodology.

MASSELINK, G. & SHORT, A.D. 1993. The effect of the tide range on beach morphodynamics and morphology: a conceptual beach model. Journal of Coastal Research, 9 (3): 785-800.

MASSAD, Faiçal. Earthworks: Basic Course in Geotechnics. 2. ed. São Paulo: Oficina de Textos Publishing House, 2003. 216p.

MATTOS, E. F. O; CERQUEIRA NETO, J. X.; SILVA, F. R.; GOMES, R. L.; OLIVEIRA,

S.M. 2005. Criteria for prioritising interventions in risk areas defined by the Slope Master Plan for the Municipality of Salvador. In: CONF. CONF. ON STABILITY

MENDES, R.M. 2001. Geotechnical mapping of the central urban area of São José do Rio Preto (SP) on a scale of 1:10,000 as a subsidy for urban planning (Master's dissertation), Postgraduate Programme in Urban Engineering, Centre for Exact Sciences and Technology, Federal University of São Carlos, 2 Vol. 245p.

MÓNICA MARIA LOPES DE SEQUEIRA AMARAL FERREIRA. Seismic risk in urban systems, 2012. Pg. 261

MORAES, D. B. The role of municipalities in environmental competence. 17/05/2007.

MOREIRA, M. E. The dynamics of the coastal systems of southern Mozambique during the last few years

30 years. Finisterra, v. 40, 79, 2005, p. 121-135

MUCHANGOS, A. dos. Mozambique: Landscapes and Natural Regions. Maputo: Author, 1999. 163 p.

NASCIMENTO, Flávio Rodrigues. Evaluation of natural resources on the island of Inhaca (2016), (Indian Ocean, Mozambique): first approach

REIGOTA, M. Challenges to school environmental education. In: JACOBI, P. et al. (eds.). Education, environment and citizenship: reflections and experiences. São Paulo: SMA, 1998. p.43-50.

RIBEIRO, J. L., Riscos Costeiros - Estratégias de prevenção, mitigação e protecção, no âmbito do planeamento de emergência e do ordenamento do território, 2010.

RICARDO Aguiar Magalhães- Companhia Energética de Minas Gerais - CEMIGVII National Symposium on Erosion Control Goiânia (GO), 03 to 06 May 2001

TONIAL, T. M. et al. Environmental diagnosis of landscape units in the north-western region of the state of rio grande do sul from 1984 to 1999. RBC - Brazilian Journal of Cartography N° 57/03, 2005. Available at: <http://www.lsie.unb.br/rbc/index.php/rbc/article/viewFile/127/110>. Accessed on: 28 Jan. 2015.Revista da Gestão Costeira Integrada 9 (1):17-37 (2009) CCELIA REGINA DE GOUVEIA SOUZA, Coastal Erosion and the Challenges of Coastal Management in Brazil

SANTOS, C.. Environmental mapping and urban territorial planning. Heritage: Leisure & Turismo, v. 6, n. 7, jul.-aug.-set./2009, p. 40-74. Available at: <http://www.unisantos.br/pos/revistapatrimonio/pdf/Artigo3 v6 n7 jul august set2009 PatrImoni o UniSantos.pdf>. Accessed on: 28 January 2015.

SORRENTINO et al, Environmental education as public policy, 2005.

PELLETIER, P. A Japan without risks? In: VEYRET, Y. (Org.) Os Riscos: o Homem como agressor e vítima do meio ambiente. São Paulo: Contexto, 2007. p. 201-220.

PULKOWNIK, A; BURCHETT, M.D; LAGINESTRA E. Review of Environmental Risk Assessments (ERAs) at Sydney Olympic Park: Method Outline and Summary, 2005.

Saket, M. (1994) Report on the updating of the exploratory national forest inventory. Maputo, FAO/UNDP: MOZ/920/13. 77p

ULRICH, BECK. The society of risk: towards a new modernity, Barcelona: Paidos.liberica, 2006

VARNES, D.J. (1984) - Landslide Hazard Zonation: Review of Principles and Practice. UNESCO Press, 56p., Paris, France (ISBN 92-3-101895-7).

Appendix

Appendix 1

Waiting for the arrival of the maritime transport link between Maputo city and the island of ka-Nyaka, the vehicles lined up in areas susceptible to coastal erosion are the vehicles of local tour operators waiting for the arrival of tourists and goods.

Source: author 2016

Appendix 2

Fence of a private property made of material from the Mangal

As the image illustrates, the use of this type of walkway is important because it guarantees the sustainability of ecosystems along areas of unconsolidated dunes

Source: Lordina Boque

So the image clearly shows the difference in the landscape, one that is more cared for, and look at figure 24, you can no longer say the same, since there is no work to take care of the ecosystems along the coastline.

These are completely different examples from the previous one, and it is recommended that the team works in coordination with the district's government structures in order to achieve the project's objectives.

Annexes i, Table 1, FCTA teachers

Name	Titilo	Position
Gustavo Sobrinho Dgedge	Prof Dr	Director of the FCTA
Suzete Lourenço Buque	Prof Dr	FCTA lecturer
Sabil Damião Mandala	Prof Dr	FCTA lecturer
António Soluda	Prof Dr	FCTA lecturer
Zacarias Ombe	Prof Dr	FCTA lecturer
José Julião da Silva	Prof Dr	FCTA lecturer

Table 2, guides and information facilitators on the island of Ka-nyaka

Name of interviewee	Institution and position	Where they come from
Sérgio Mapanga	UEM- (EBM), Employee	Native and resident
Fernando Nhaca	Tour operator	Native and resident
khenssane	Tour Operator	Native and resident
Rita Erica bar	Trader	Native and resident
Angelo Manguele	Tour guide/entrepreneur	Native and resident
Paulo Manguel	CMCM Engineer	840600872
Araújo	DMK Councillor	864849150
Mário Mazive	Sculptor	
Jeremias Manussa	Director/Tourism CMCM	827907568

Printed by Books on Demand GmbH, Norderstedt / Germany